RAISED-BED GARDENING FOR BEGINNERS

By

DR. ANITA BRAINERD

CHAPTER ONE 5

INTRODUCTION TO RAISED BED GARDENING ... 5

1. What is raised bed gardening? 5

2. Benefits of raised bed gardening 5

3. Why choose raised beds over traditional gardening methods? 9

4. Materials needed for building raised beds 11

CHAPTER TWO 13

PLANNING YOUR RAISED BED GARDEN 13

1. Choosing the right location for your raised beds 13

2. Determining the size and shape of your raised beds. 16

3. Sunlight and drainage considerations 19

4. Sketching out your garden layout 21

CHAPTER THREE 23

BUILDING YOUR RAISED BEDS 23

1. Selecting materials (wood, bricks, concrete blocks, etc.) 23

2. Step-by-step instructions for constructing raised beds 25

3. Options for adding legs or benches to raised beds.... 28

4. Tips for ensuring proper drainage and stability 29

CHAPTER FOUR 31

SOIL PREPARATION AND FILLING YOUR RAISED BEDS 31

1. Soil selection and quality considerations 31

2. Amending soil for optimal plant growth 34

3. Filling your raised beds with soil mixtures 36

4. Techniques for leveling and smoothing soil surfaces 37

CHAPTER FIVE 39

PLANTING IN RAISED BEDS 39

1. Choosing plants suited for raised bed gardening 39

2. Spacing and layout tips for different types of plants 41

3. Companion planting strategies for maximizing space and yield 43

CHAPTER SIX 47

MAINTENANCE AND CARE 47

1. Watering schedules and techniques for raised beds .. 47

2. Mulching to conserve moisture and suppress weeds 50

3. Fertilizing and nourishing your plants 52

4. Pest and disease management in raised bed gardens 54

CHAPTER SEVEN 57

HARVESTING AND BEYOND 57

1. Signs of readiness for harvesting various crops 57

2. Harvesting techniques to avoid damage to plants 60

3. Post-harvest care and cleanup 63

4. Planning for future seasons and crop rotation 66

CHAPTER ONE

INTRODUCTION TO RAISED BED GARDENING

1. What is raised bed gardening?

Raised bed gardening is a method of gardening where soil is contained within a framed structure above ground level, rather than directly in the ground. The frame can be made of various materials like wood, bricks, or stone. This technique offers several advantages, including better soil drainage, improved soil temperature control, and easier weed and pest management. Raised beds also provide a more accessible gardening space, reducing strain on the gardener's back and knees. Overall, raised bed gardening is a practical and efficient way to grow plants in a controlled environment.

2. Benefits of raised bed gardening

➢ **Improved Drainage:** Raised beds generally have better drainage compared to traditional in-ground gardens. The elevated nature of the beds allows excess water to drain away more effectively, preventing waterlogging and root rot.

➢ **Better Soil Quality:** Gardeners have greater control over the soil composition in raised beds. They can fill the beds with high-quality soil, rich in organic matter, and tailor it to the specific needs of the plants. This results in healthier plants with improved growth.

➢ **Warmer Soil Temperature:** Raised beds tend to warm up faster in the spring, allowing for earlier planting and extended growing seasons. Warmer soil temperatures promote faster germination and growth of plants.

➢ **Easier Access:** The raised height of the beds makes it more comfortable for gardeners to access and maintain their plants. This is particularly beneficial for individuals with mobility issues, as it reduces the need for bending or kneeling.

➢ **Weed Control:** Raised beds make it easier to control weeds. The contained space and controlled soil environment make it more challenging for weeds to establish and spread. Mulching the surface can further suppress weed growth.

➢ **Pest Management:** Raised beds can help in pest control. The defined borders of the beds make it easier to implement protective measures like row covers, nets, or barriers to keep pests away from plants.

➢ **Aesthetically Pleasing:** Raised beds can enhance the visual appeal of a garden. The neat, organized layout of raised beds adds structure and order to the garden, making it more aesthetically pleasing.

➢ **Water Conservation:** Since the beds are elevated, water is less likely to run off, and it can be more efficiently directed to the root zone of plants. This helps in water conservation and reduces the overall water requirements for the garden.

➢ **Customization and Design:** Gardeners have the flexibility to design raised beds in various shapes and sizes, allowing for creative and customized layouts. This customization can be aesthetically pleasing and functional.

➢ **Reduced Compaction:** Since gardeners can access their plants without walking on the growing area, the soil in raised beds is less likely to become compacted. This promotes better root growth and overall plant health.

➢ **Optimized Plant Spacing:** Gardeners can plan and optimize the spacing of plants more efficiently in raised beds. This ensures that each plant has adequate space for growth and reduces competition for nutrients.

3. Why choose raised beds over traditional gardening methods?

Raised beds offer several advantages over traditional gardening methods, making them a popular choice for many gardeners. Firstly, raised beds provide better soil drainage, preventing waterlogging and ensuring optimal root health. This enhanced drainage can be especially beneficial in areas with heavy rainfall or clay soils.

Secondly, raised beds allow for better soil quality control. Gardeners can fill raised beds with high-quality soil, compost, and amendments, creating an optimal growing environment for plants. This is particularly advantageous in areas where the native soil may be poor or contaminated.

Additionally, raised beds offer improved weed control. The elevated structure makes it easier to spot and remove weeds, reducing competition for nutrients and water among plants.

Furthermore, raised beds can extend the growing season. The soil in raised beds tends to warm up faster in the spring, allowing for earlier planting and longer growing periods. This can be especially advantageous for gardeners in cooler climates.

Raised beds also provide better accessibility for gardeners with mobility issues. The raised height reduces the need for bending and kneeling, making gardening tasks more comfortable and accessible.

Lastly, raised beds can be aesthetically pleasing and can help organize garden spaces more effectively. They can be constructed in various shapes, sizes, and materials, allowing for creative design possibilities that enhance the overall look of the garden.

Overall, the benefits of raised beds, including improved drainage, soil quality control, weed control, extended growing seasons, accessibility, and aesthetics, make them a preferred choice for many gardeners seeking efficient and productive gardening solutions.

4. Materials needed for building raised beds

- Lumber: Choose rot-resistant lumber such as cedar or redwood for longevity. You'll need enough boards to create the sides of your raised beds. The dimensions will vary depending on the size of your beds, but typically 2x6 or 2x8 boards work well.
- Fasteners: Opt for corrosion-resistant screws or nails to secure the boards together. Screws provide more strength and durability than nails.
- Corner Braces (optional): Corner braces can reinforce the joints of the raised beds, providing extra stability and preventing warping over time.
- Soil: High-quality soil is crucial for the success of your raised bed garden. Purchase a mix of topsoil, compost, and other organic matter to fill your beds. The amount needed depends on the size and depth of your beds.

- Landscape Fabric (optional): Some gardeners choose to line the bottom of their raised beds with landscape fabric to prevent weeds from growing up into the bed. This is particularly useful if you're placing the beds over grass or weeds.
- Tools: You'll need basic carpentry tools such as a saw, drill or hammer, measuring tape, and level to assemble the raised beds.
- Safety Gear: Don't forget safety gear like gloves and goggles to protect yourself while cutting and assembling the lumber.
- Watering System (optional): Depending on your preference, you may want to install a drip irrigation system or soaker hoses to ensure consistent watering for your plants.
- Mulch (optional): Mulch helps retain moisture in the soil, suppresses weeds, and regulates soil temperature. Consider adding a layer of mulch to the surface of your raised beds once they're filled with soil.

CHAPTER TWO

PLANNING YOUR RAISED BED GARDEN

1. Choosing the right location for your raised beds

➢ **Sunlight:** Ensure your raised beds receive adequate sunlight. Most vegetables and fruits require at least 6-8 hours of direct sunlight daily for optimal growth. Observe the sunlight patterns in your yard throughout the day to determine the sunniest spots.

➢ **Drainage:** Proper drainage is essential to prevent waterlogging, which can lead to root rot and other plant diseases. Avoid low-lying areas where water tends to accumulate. Additionally, consider elevating your raised beds slightly or incorporating drainage materials such as gravel at the base.

➢ **Accessibility:** Choose a location that is easily accessible for planting, watering, weeding, and harvesting. This is especially important if you have physical limitations or mobility issues. Placing raised beds closer to your home or pathways can make maintenance tasks more convenient.

➢ **Soil quality:** Assess the quality of the soil in your chosen location. Avoid areas with compacted or poor-quality soil. Raised beds allow you to control the soil composition more effectively, but starting with a decent soil base can save you time and effort in amending the soil later on.

➢ **Protection from wind and pests:** Consider the exposure of your raised beds to wind and potential pests. While some airflow is beneficial for preventing diseases, excessive wind can damage delicate plants. Installing windbreaks or placing raised beds in sheltered areas can mitigate this risk. Similarly, if you have persistent pest issues, choose a location that is easier to protect with barriers or netting.

- ➢ **Microclimates:** Be aware of microclimates in your yard. Factors such as buildings, trees, or slopes can create variations in temperature and sunlight. Understand these microclimates to optimize the growing conditions for different plants.

- ➢ **Aesthetics:** Consider the visual appeal of your garden when choosing the location for your raised beds. Select a spot that complements the overall landscape and enhances the beauty of your outdoor space.

2. Determining the size and shape of your raised beds

➤ **Assess Available Space:** Measure the area where you plan to install the raised beds. Consider factors like sunlight exposure, water accessibility, and any potential obstructions.

➤ **Consider Accessibility:** Ensure your raised beds are easily accessible from all sides to make planting, weeding, and harvesting more convenient. If you have limited mobility, consider making the beds narrower to reach across comfortably.

➤ **Choose a Shape:** Raised beds can be rectangular, square, circular, or even irregularly shaped. Rectangular or square beds are common because they maximize space and are easier to manage, but feel free to get creative with your design.

➢ **Determine Size:** The dimensions of your raised beds will depend on your space and gardening needs. Aim for a width of 3 to 4 feet, as this allows you to reach the center from either side without stepping into the bed, which can compact the soil. Length can vary depending on available space and aesthetic preferences.

➢ **Height:** Raised beds typically range from 6 to 12 inches in height, but this can vary based on factors like the depth of your soil, the root depth of your plants, and your own comfort level. Taller beds can be easier to work in, especially for those with back problems.

➢ **Materials:** Consider the materials you'll use to construct your raised beds. Common options include wood, concrete blocks, bricks, or even recycled materials like old tires or pallets.

➤ **Spacing:** Leave enough space between raised beds for pathways to walk and work comfortably. A width of 2 to 3 feet between beds is usually sufficient.

➤ **Water Drainage:** Ensure proper drainage by incorporating drainage holes or gravel at the bottom of the raised beds, especially if you're using impermeable materials like wood or plastic.

➤ **Plan for Future Expansion:** If you anticipate expanding your gardening efforts in the future, leave space for additional raised beds or consider building modular beds that can be easily added onto later.

➤ **Sketch Your Design:** Once you've considered all these factors, sketch out your design to visualize how your raised beds will fit into your garden space.

3. Sunlight and drainage considerations

Sunlight is essential for photosynthesis, the process by which plants convert light energy into chemical energy to fuel their growth. When planning raised beds, ensure they receive adequate sunlight according to the needs of the plants you intend to grow. Most vegetables and flowers require at least 6-8 hours of direct sunlight daily for optimal growth. Choose a location for your raised beds that receives ample sunlight throughout the day, preferably facing south or southwest for maximum exposure.

Proper drainage is equally important to prevent waterlogging, which can lead to root rot and other problems. Raised beds typically have better drainage compared to traditional garden beds due to their elevated nature. However, it's essential to ensure that excess water can drain away efficiently. Use well-draining soil mixes in your raised beds, consisting of a combination of compost, sand, and organic matter. Avoid compacting the soil, as this can impede drainage.

To further enhance drainage, consider incorporating drainage features into your raised beds, such as perforated drainage pipes or gravel layers at the bottom. Additionally, avoid overwatering your raised beds, as excess water can accumulate and cause drainage issues. Water the plants deeply but infrequently, allowing the soil to dry out slightly between watering sessions.

Regularly monitor the moisture levels in your raised beds to ensure they remain within the optimal range for your plants. Adjust watering frequency as needed based on weather conditions and the specific requirements of the plants. By providing adequate sunlight and ensuring proper drainage, you can create an ideal growing environment for your raised-bed garden, promoting healthy plant growth and abundant harvests.

4. Sketching out your garden layout

- ➤ **Measure Your Space:** Begin by measuring the dimensions of your garden area. This will help you accurately represent the size of your garden on paper.

- ➤ **Identify Key Features:** Take note of any existing features in your garden such as trees, shrubs, pathways, or structures like sheds or fences. These will be important elements to include in your sketch.

- ➤ **Set Goals:** Decide what you want to achieve with your garden layout. Are you looking to create a relaxing retreat, a vibrant flower garden, or a productive vegetable patch?

- ➤ **Design Zones:** Divide your garden into different zones based on your goals and the needs of the plants you want to include. For example, you might have a seating area, a vegetable garden, a flower bed, and a lawn area.

- ➤ **Create a Rough Sketch:** Start by drawing a rough outline of your garden space on a piece of paper.

Use a pencil so you can easily make changes as you go along.

➢ **Add Features:** Begin adding the key features you identified earlier to your sketch. Use simple shapes and symbols to represent different elements such as circles for trees, rectangles for beds, and lines for pathways.

➢ **Consider Scale:** Make sure your sketch accurately represents the scale of your garden. You can use a scale ruler to help ensure your measurements are proportional.

➢ **Add Detail:** Once you have the basic layout in place, you can start adding more detail to your sketch. Include plant names, furniture, and any other features you want to include in your garden.

➢ **Review and Revise:** Step back and take a look at your sketch. Make any necessary revisions to ensure everything flows well and meets your goals.

➢ **Finalize Your Plan:** Once you're happy with your garden layout sketch, you can use it as a guide to start planning and planting your garden in real life!

CHAPTER THREE

BUILDING YOUR RAISED BEDS

1. Selecting materials (wood, bricks, concrete blocks, etc.)

➤ **Wood:** Wood is a popular choice due to its natural look and ease of construction. Cedar and redwood are naturally rot-resistant and can last for years. However, untreated wood may degrade over time, so consider using rot-resistant varieties or lining the bed with plastic to prolong its lifespan. Avoid pressure-treated wood, as it may contain chemicals that can leach into the soil.

➤ **Bricks:** Bricks offer a classic, timeless look and provide excellent stability for raised beds. They are durable and can withstand weathering over time. However, they can be more labor-intensive to install compared to other materials, as they need to be stacked and mortared together securely.

➢ **Concrete blocks:** Concrete blocks are sturdy and long-lasting, making them a popular choice for raised beds. They are easy to stack and provide a clean, modern look. However, they can be heavy and may require machinery to install, which can add to the overall cost.

➢ **Composite materials:** Composite materials, such as recycled plastic lumber or composite decking boards, offer the durability of wood without the maintenance. They are resistant to rot, insects, and weathering, making them a low-maintenance option for raised beds. However, they can be more expensive upfront compared to other materials.

➢ **Metal:** Metal raised beds, typically made from galvanized steel or corten steel, offer a sleek and contemporary look. They are extremely durable and resistant to rot and pests. However, they can be costly and may heat up more quickly than other materials, which can affect soil temperature.

2. Step-by-step instructions for constructing raised beds

MATERIAL NEEDED

i. Lumber (cedar is often preferred for its durability, but any untreated wood will work)
ii. Screws or nails
iii. Drill or hammer
iv. Level
v. Soil
vi. Compost (optional)
vii. Plants or seeds

Step 1: Choose a Location

- Select a flat area with good sunlight exposure for your raised bed.

➤ **Step 2: Determine Size and Shape**

- Decide on the dimensions and shape of your raised bed. Common sizes are 4x8 feet or 3x6 feet, but you can adjust according to your space and needs.

➤ **Step 3: Gather Materials**

- Purchase or gather the necessary materials, including lumber and screws or nails.

➤ **Step 4: Cut Lumber**

- Use a saw to cut the lumber to the desired lengths according to your chosen dimensions.

➤ **Step 5: Assemble the Frame**

- Lay out the cut lumber in the shape of your raised bed.

- Use screws or nails to connect the corners of the frame. Make sure the corners are square using a carpenter's square.

➤ **Step 6: Level the Bed**

- Use a level to ensure that the frame is even and level on the ground. Adjust as necessary.

Step 7: Place the Raised Bed

- Position the assembled frame in the desired location in your garden.

➤ **Step 8: Fill with Soil**

- Fill the raised bed with a mixture of topsoil and compost. You can purchase pre-mixed soil or create your own blend.
- Make sure the soil is evenly distributed and level within the raised bed.

➤ **Step 9: Plant**

- Once the raised bed is filled with soil, you can begin planting your desired plants or seeds according to your garden plan.
- Water the plants thoroughly after planting.

➤ **Step 10: Maintain**

- Regularly water and maintain your raised bed garden as needed, including weeding and adding compost or fertilizer.

Following these steps will help you construct a functional and productive raised bed for your garden!

3. Options for adding legs or benches to raised beds

➢ **Wooden Legs:** You can attach wooden legs to the corners of the raised bed frame, ensuring they are sturdy and secure. This is a simple DIY option that allows you to customize the height of the bed according to your preference.

➢ **Metal Brackets:** Metal brackets can be used to attach legs to the raised bed frame. These brackets provide strong support and stability, and they are relatively easy to install.

➢ **Cinder Blocks:** Stack cinder blocks at each corner of the raised bed to create a sturdy base. You can adjust the height by adding or removing blocks as needed. This option also provides additional planting space on top of the blocks.

- ➢ **Built-in Benches:** Instead of adding separate legs, you can incorporate benches into the design of the raised bed. This not only provides seating but also adds aesthetic appeal to your garden.
- ➢ **Foldable Legs:** If you prefer versatility, consider adding foldable legs to the raised bed. This allows you to collapse the legs for easy storage or transport when not in use.
- ➢ **Recycled Materials:** Get creative by using recycled materials such as old pallets or reclaimed lumber to construct legs or benches for your raised beds. This not only reduces waste but also adds a unique touch to your garden design.

4. Tips for ensuring proper drainage and stability

- ➢ **Use the right soil mix:** A well-draining soil mix is essential. Combine equal parts of garden soil, compost, and coarse sand or perlite to promote good drainage while retaining moisture.
- ➢ Install drainage material: Place a layer of gravel or small rocks at the bottom of the raised bed to

enhance drainage. This prevents water from pooling at the bottom of the bed, which can lead to root rot.

➤ Add drainage holes: If your raised bed doesn't have built-in drainage, consider drilling holes in the bottom to allow excess water to escape.

➤ Avoid compacting the soil: Regularly aerate the soil by gently loosening it with a hand tool or garden fork. Compacted soil inhibits drainage and root growth.

➤ Choose sturdy materials: Opt for durable materials like cedar or pressure-treated lumber for the frame of the raised bed. Avoid using materials that may degrade quickly or leach harmful substances into the soil.

➤ Consider bed dimensions: Keep the width of the raised bed manageable (no wider than 4 feet) to ensure easy access from all sides without compacting the soil in the middle.

➤ Monitor moisture levels: Keep an eye on moisture levels in the soil. Water thoroughly when the top inch of soil feels dry to the touch, but avoid overwatering, which can lead to drainage issues and root rot.

CHAPTER FOUR

SOIL PREPARATION AND FILLING YOUR RAISED BEDS

1. Soil selection and quality considerations

> **Texture and Structure:** Choose a well-draining soil with good texture and structure. A balanced mix of sand, silt, and clay is ideal. This promotes proper aeration, root development, and water drainage.

> **pH Level:** Check and adjust the pH of the soil to suit the specific needs of the plants you intend to grow. Most vegetables prefer slightly acidic to neutral pH levels (around 6.0 to 7.0).

> **Organic Matter:** Incorporate organic matter like compost or well-rotted manure into the soil. This improves fertility, enhances water retention, and provides essential nutrients for plant growth.

➤ **Nutrient Content:** Conduct a soil test to determine the nutrient levels in the soil. Based on the results, add fertilizers or amendments to ensure the soil has adequate amounts of essential nutrients for plant growth.

➤ **Water Retention:** While drainage is important, the soil should also retain enough moisture for the plants. Organic matter helps in water retention, and you can also use mulch to reduce evaporation.

➤ **Avoid Compaction:** Raised beds can sometimes suffer from soil compaction. Regularly turning and loosening the soil helps prevent compaction and promotes better root penetration.

➤ **Contaminant-Free Soil:** Ensure the soil is free from contaminants such as heavy metals, pesticides, or other pollutants. Contaminated soil can negatively affect plant health and pose risks to human consumption.

➢ **Local Climate Considerations:** Consider the local climate conditions when selecting soil. Certain regions may require specific soil characteristics to cope with temperature extremes, rainfall patterns, or other climate-related factors.

➢ **Consistent Soil Depth:** Aim for a consistent soil depth in the raised bed to allow for uniform root development. This is especially important for crops with deeper root systems.

➢ **Mulching:** Mulching the surface of the soil can help regulate temperature, reduce water evaporation, and control weed growth. Use organic mulches like straw or wood chips for added benefits.

2. Amending soil for optimal plant growth

➢ **Assess Your Soil:** Before amending the soil, it's essential to understand its current composition. You can conduct a soil test to determine pH levels, nutrient deficiencies, and texture.

➢ **Add Organic Matter:** Organic matter, such as compost, well-rotted manure, or leaf mold, is crucial for improving soil structure, water retention, and nutrient availability. Aim to add a few inches of organic matter to the raised bed and mix it thoroughly with the existing soil.

➢ **Balance Nutrients:** Based on your soil test results, you may need to add specific nutrients to achieve optimal levels. This could include adding fertilizers or organic amendments rich in nitrogen, phosphorus, and potassium, as well as micronutrients like calcium, magnesium, and sulfur.

➤ **Adjust pH:** If the soil pH is too high or too low for your plants' needs, you can adjust it by incorporating materials such as lime to raise pH or elemental sulfur to lower pH. Again, this should be based on the results of your soil test.

➤ **Mulch:** Once the soil is amended, consider applying a layer of mulch on top. Mulch helps conserve moisture, suppress weeds, and regulate soil temperature.

➤ **Monitor and Maintain:** Regularly monitor the soil moisture, nutrient levels, and plant health throughout the growing season. You may need to top-dress with compost or add supplemental fertilizer if necessary.

➤ **Rotate Crops:** To prevent nutrient depletion and disease buildup, practice crop rotation in your raised beds. Rotate plants of different families each season to optimize soil health and fertility.

3. Filling your raised beds with soil mixtures

➢ **Base Layer:** Start with a layer of coarse materials like gravel or small rocks at the bottom of the raised bed to improve drainage and prevent waterlogging.

➢ **Filler Material:** Next, add a layer of filler material like wood chips or shredded leaves. This helps with aeration and moisture retention, promoting healthy root growth. **Topsoil:** Follow up with a layer of quality topsoil. Look for soil that is rich in organic matter and nutrients. This layer provides the main growing medium for your plants.

➢ **Compost:** Mix in a generous amount of compost with the topsoil layer. Compost adds essential nutrients to the soil and improves its structure, helping plants thrive.

➢ **Additional Amendments:** Depending on your plants' needs and your soil's composition, you may want to add additional amendments like perlite for improved drainage or lime to adjust pH levels.

➢ **Mix Thoroughly:** Once all layers are added, mix them thoroughly to ensure even distribution of materials and nutrients throughout the raised bed.

➢ **Mulch:** Finally, top off the soil with a layer of organic mulch like straw or shredded bark. Mulch helps retain moisture, suppress weeds, and regulate soil temperature.

4. Techniques for leveling and smoothing soil surfaces

➢ **Hand Raking:** Use a garden rake to evenly distribute soil across the raised bed surface. This method allows for precise control and is suitable for small areas.

➢ **Dragging a Board:** Place a long, straight board (such as a 2x4) across the raised bed and drag it back and forth to level the soil. This method helps to create a uniform surface and is effective for larger areas.

➢ **Garden Hoe:** A garden hoe can be used to break up clumps of soil and smooth out the surface. Use the flat side of the hoe to gently scrape and level the soil.

- **Soil Tamper:** A soil tamper or hand tamper can be used to compact soil and create a smooth, firm surface. Simply press the tamper down onto the soil and walk across the raised bed to even out the surface.
- **Watering and Compacting:** After leveling the soil, lightly water the raised bed to help settle the soil particles. Use a roller or the back of a shovel to gently compact the soil surface, creating a smooth finish.
- **Layering:** If your raised bed soil is uneven due to different types of materials (such as compost, topsoil, and amendments), consider layering the materials in thin, even layers and then mixing them together with a garden fork or shovel.
- **Use a Level:** For precise leveling, you can use a carpenter's level to ensure that the soil surface is flat and even. Place the level across the raised bed in different directions to check for any uneven areas.
- **Add More Soil:** If certain areas of the raised bed are significantly lower than others, you may need to add more soil to achieve a level surface. Spread the additional soil evenly across the bed and then use one of the above techniques to level and smooth it out.

CHAPTER FIVE

PLANTING IN RAISED BEDS

1. Choosing plants suited for raised bed gardening

➢ **Vegetables:**

- Tomatoes: Choose compact or determinate varieties that don't sprawl excessively.

- Peppers: Both sweet and hot peppers thrive in raised beds and don't require much space.

- Lettuce: Fast-growing and shallow-rooted, perfect for raised beds with limited depth.

- Spinach: Another leafy green that grows well in raised beds and can be harvested multiple times.

- Radishes: Quick-growing and don't need deep soil, making them ideal for raised beds.

➢ **Herbs:**

- Basil: Thrives in raised beds, especially in warmer climates, and adds flavor to many dishes.

- Parsley: Grows well in containers and raised beds, providing a steady supply of fresh herbs.

- Chives: Perennial herbs that are easy to grow and do well in raised beds.

- Thyme: Low-growing and drought-tolerant, suitable for raised beds with good drainage.

- Rosemary: Prefers well-drained soil, making it a good choice for raised beds.

➢ **Fruits:**

- Strawberries: Compact plants that produce well in raised beds, especially if you choose everbearing varieties.

- Blueberries: Acidic soil-loving plants that can thrive in raised beds with the right soil mix.

- Dwarf fruit trees: Consider dwarf varieties of apples, peaches, or cherries for larger raised beds.

➢ **Flowers:**

- Marigolds: These colorful flowers help deter pests and add beauty to raised beds.
- Nasturtiums: Edible flowers that attract pollinators and can be used as a trap crop for aphids.
- Zinnias: Easy to grow and provide long-lasting blooms, adding color to raised beds.

➢ **Perennials:**

- Asparagus: Deep-rooted perennial vegetable that can thrive in raised beds for many years.
- Rhubarb: Another perennial vegetable that does well in raised beds with rich, well-draining soil.

2. Spacing and layout tips for different types of plants

➢ **Vegetables:**

- Row Planting: Arrange vegetables in rows with adequate spacing between each row to allow for easy access and airflow.

- Square Foot Gardening: Divide the raised bed into square-foot sections and plant different vegetables according to their recommended spacing per square foot.

- Companion Planting: Consider companion planting to maximize space and enhance growth. For example, plant tall or vining crops like tomatoes or cucumbers beside shorter ones like lettuce or radishes to optimize space.

➢ **Herbs:**

- Grouping: Plant herbs with similar water and sunlight requirements together in clusters or groups.

- Edging: Use herbs with trailing or low-growing habits to edge the raised bed, providing a neat and attractive border.

➢ Flowers:

- Color Coordination: Plan the layout to create visual interest by mixing flowers with complementary colors and varying heights.

- Thrillers, Fillers, and Spillers: Arrange flowers according to the "thriller, filler, spiller" concept, where taller plants (thrillers) are placed in the center or back, medium-sized plants (fillers) fill the middle, and trailing or cascading plants (spillers) drape over the edges.

> **Perennials:**

- Spacing for Growth: Allow sufficient space between perennial plants to accommodate their mature size and prevent overcrowding.

- Grouping: Plant perennials in groups or clusters, considering their bloom times, colors, and textures for a visually appealing display.

> **Fruit Trees and Bushes:**

- Espalier or Trellis: Train fruit trees or bushes against a trellis or espalier them to save space and promote airflow.

- Dwarf Varieties: Opt for dwarf or semi-dwarf varieties of fruit trees and bushes to fit within the confines of the raised bed while still providing a fruitful harvest.

3. Companion planting strategies for maximizing space and yield

> **Tall and Short Plants:** Plant taller plants like tomatoes, corn, or trellised cucumbers on the north

side of the bed to prevent shading shorter plants. This ensures all plants receive adequate sunlight.

➢ **Three Sisters:** The Three Sisters planting technique involves planting corn, beans, and squash together. Corn provides a structure for the beans to climb, beans fix nitrogen in the soil, benefiting the corn and squash, while squash shades the soil, conserving moisture and suppressing weeds.

➢ **Interplanting:** Interplant fast-growing crops like lettuce, radishes, or spinach between slower-growing plants like tomatoes or peppers. This maximizes space and allows you to harvest multiple crops from the same area.

➢ **Complementary Plants:** Pair plants that have complementary growth habits or nutrient needs. For example, plant shallow-rooted crops like lettuce with deep-rooted crops like carrots to avoid competition for nutrients.

> **Beneficial Insect Attractors:** Include plants that attract beneficial insects like ladybugs, lacewings, or predatory wasps. For example, planting marigolds or dill can attract beneficial insects that prey on pests like aphids or caterpillars.

> **Trap Crops:** Plant trap crops like nasturtiums or radishes to attract pests away from main crops. This helps protect your main crops from pest damage.

> **Succession Planting:** Follow an early harvest crop with a second crop that matures later in the season. For example, after harvesting lettuce, plant heat-loving crops like peppers or eggplants.

> **Companion Herbs:** Integrate herbs like basil, thyme, or cilantro among your vegetable crops.

These herbs can improve the flavor of nearby vegetables and deter pests.

➤ **Vertical Gardening:** Utilize trellises or vertical structures to grow vining plants like cucumbers, peas, or beans, which can free up ground space for other crops.

➤ **Crop Rotation:** Plan your planting layout to rotate crops each season to prevent the buildup of pests and diseases. Rotate heavy feeders with nitrogen-fixing plants or cover crops to replenish soil nutrients.

CHAPTER SIX

MAINTENANCE AND CARE

1. Watering schedules and techniques for raised beds

> **Mulching:** Apply a layer of organic mulch such as straw, wood chips, or compost to the surface of the soil. Mulch helps retain moisture in the soil, reduces evaporation, and prevents weed growth. This can significantly reduce the frequency of watering.

> **Drip Irrigation:** Install a drip irrigation system in your raised beds. Drip irrigation delivers water directly to the base of plants, minimizing water loss through evaporation and ensuring that water reaches the roots where it's needed most. Use a timer to automate watering schedules for consistent moisture levels.

➢ **Watering Frequency:** The frequency of watering will depend on various factors such as the type of soil, weather conditions, and the specific needs of your plants. In general, raised beds may require more frequent watering than traditional garden beds because they tend to drain more quickly. Check the moisture level of the soil regularly by sticking your finger into the soil about an inch deep. If it feels dry, it's time to water.

➢ **Morning Watering:** Water your raised beds in the morning to minimize evaporation loss. Watering in the morning allows plants to absorb moisture before the heat of the day, reducing stress and promoting healthy growth.

➢ **Deep Watering:** When you water, do so deeply and thoroughly to encourage deep root growth. Shallow watering can lead to shallow roots, making plants more susceptible to drought stress. Water until the soil is evenly moist to a depth of at least 6 inches.

> **Watering Techniques:** Avoid overhead watering if possible, as it can promote fungal diseases and water wastage due to evaporation. Instead, aim to water the soil directly at the base of plants. For smaller raised beds, watering cans or handheld hoses with a gentle spray attachment can be effective. For larger beds, consider soaker hoses or drip irrigation systems.

> Adjust Based on Weather: Be prepared to adjust your watering schedule based on weather conditions. During hot and dry periods, you may need to water more frequently, while rainy periods may require less frequent watering.

> Observation and Adjustment: Pay attention to how your plants respond to watering. Wilting or

yellowing leaves can indicate both overwatering and underwatering. Adjust your watering schedule accordingly to meet the specific needs of your plants.

2. Mulching to conserve moisture and suppress weeds

ulching in raised beds is a highly effective practice for conserving moisture and suppressing weeds. By applying a layer of organic or inorganic material to the soil surface, you create a protective barrier that offers several benefits.

Firstly, mulch helps to conserve moisture by reducing evaporation from the soil. This is particularly crucial in raised beds, where the soil tends to dry out more quickly due to increased exposure. The mulch acts as a shield, preventing water from escaping into the atmosphere and maintaining a more stable soil moisture level.

Secondly, mulching suppresses weeds by blocking sunlight and inhibiting weed germination. Weeds struggle to penetrate through the mulch layer, reducing the competition for water and nutrients with your desired plants. This is especially important in raised beds where space is limited, and weed control is essential for plant health.

When choosing mulch for raised beds, organic options like straw, shredded leaves, or compost offer additional benefits. They break down over time, adding organic matter to the soil, improving its structure, and enhancing nutrient content. Inorganic mulches such as plastic or landscape fabric provide longer-lasting weed suppression and moisture retention but do not contribute to soil fertility.

In conclusion, mulching in raised beds is a simple yet powerful technique to ensure optimal moisture levels and minimize weed interference, promoting a healthier and more productive gardening environment.

3. Fertilizing and nourishing your plants

> Soil Preparation: Begin by ensuring your raised bed has well-draining soil rich in organic matter. Mix in compost or aged manure to improve soil structure and fertility.

> **Choose the Right Fertilizer:** Select a balanced fertilizer or organic amendments suited to the specific needs of your plants. Consider factors like nitrogen (N), phosphorus (P), and potassium (K) levels.

> **Timing:** Apply fertilizers at the right time according to the needs of your plants. Generally, fertilizing at the beginning of the growing season and periodically throughout ensures steady nutrient availability.

➢ **Application Methods:** Distribute fertilizer evenly across the raised bed surface, avoiding direct contact with plant stems to prevent burning. Incorporate granular fertilizers into the soil before planting, or use liquid fertilizers for quick absorption.

➢ **Mulching:** Mulch the surface of the raised bed with organic materials like straw or wood chips to conserve moisture and regulate soil temperature. As the mulch breaks down, it enriches the soil with nutrients.

➢ **Regular Monitoring:** Keep an eye on plant growth and health to identify any signs of nutrient deficiencies or excesses. Adjust fertilizer applications accordingly to maintain a balanced nutrient profile.

➢ **Watering:** Ensure plants receive adequate water after fertilizing to facilitate nutrient uptake.

However, avoid overwatering, as it can leach nutrients from the soil.

➢ **Crop Rotation:** Practice crop rotation in your raised beds to prevent nutrient depletion and minimize the buildup of pests and diseases associated with specific plant families.

4. Pest and disease management in raised bed gardens

➢ **Start with Healthy Soil:** Healthy soil supports healthy plants that are more resistant to pests and diseases. Use high-quality soil mixtures rich in organic matter and well-draining.

➢ **Practice Crop Rotation:** Avoid planting the same crops in the same raised beds year after year. Rotate crops to disrupt pest and disease cycles and prevent buildup in the soil.

➢ Companion Planting: Some plants naturally repel pests or attract beneficial insects. Consider interplanting crops with companion plants that help deter pests or attract predatory insects.

➢ **Use Physical Barriers:** Install barriers such as row covers or netting to protect plants from pests like birds, rodents, and insects. This is particularly useful for crops vulnerable to aerial attacks.

➢ **Mulching:** Apply organic mulch around plants to suppress weed growth, conserve moisture, and reduce the spread of soil-borne diseases.

➢ **Regular Inspection:** Check your raised beds regularly for signs of pests or diseases. Look for chewed leaves, wilting, discoloration, or other symptoms of stress.

➢ **Integrated Pest Management (IPM):** Adopt an IPM approach, which combines various tactics such as cultural, biological, and chemical methods to manage pests while minimizing environmental impact. Only resort to chemical controls when absolutely necessary and use them sparingly and according to label instructions.

➢ **Handpick Pests:** For smaller infestations, manually remove pests like caterpillars, beetles, or snails from plants. Wear gloves and dispose of pests away from the garden to prevent reinfestation.

- ➤ **Beneficial Insects:** Encourage natural predators like ladybugs, lacewings, and parasitic wasps by planting flowers that attract them or by purchasing and releasing them into your garden.
- ➤ **Disease Prevention:** Minimize the spread of diseases by avoiding overhead watering, spacing plants properly for good air circulation, and sanitizing tools regularly to prevent contamination.
- ➤ **Crop Selection:** Choose disease-resistant varieties whenever possible. Many plant varieties are bred to resist common diseases, reducing the need for chemical controls.
- ➤ **Monitor Weather Conditions:** Be aware of weather patterns that can contribute to pest and disease outbreaks, such as prolonged periods of rain or humidity, which can favor certain pathogens.
- ➤ **Proper Plant Spacing:** Avoid overcrowding plants in raised beds, as this can create favorable conditions for diseases by restricting airflow and promoting humidity.

CHAPTER SEVEN

HARVESTING AND BEYOND

1. Signs of readiness for harvesting various crops

> **Root Vegetables (Carrots, Radishes, Beets):**

- Look for the top of the root protruding slightly above the soil.

- Gently pull aside the soil to check the size and color of the root.

- Carrots and radishes typically mature in 30-60 days, depending on the variety.

> **Leafy Greens (Lettuce, Spinach, Kale):**

- Harvest leaves when they are young and tender for the best flavor.

- For loose-leaf varieties, you can harvest outer leaves as needed, allowing the inner leaves to continue growing.

- Harvest entire heads of lettuce or spinach by cutting them just above the soil level.

➢ **Herbs (Basil, Parsley, Cilantro):**

- Harvest herbs as soon as they reach a size where you can use them.

- Regular harvesting encourages new growth and helps to prevent plants from bolting (going to seed).

➢ **Tomatoes:**

- Look for firm, fully colored fruits. The color depends on the variety (red, yellow, orange, etc.).

- Tomatoes should be slightly soft when gently squeezed.

- For determinate varieties, most fruits ripen at the same time. For indeterminate varieties, fruits ripen gradually over the season.

- **Peppers:**

- Peppers can be harvested at any stage, but they are typically harvested when they reach full size and color.

- Firmness and color are good indicators of readiness.

➢ **Cucumbers:**

- Harvest cucumbers when they are firm, crisp, and uniformly green.

- Check for the appropriate size for the variety you are growing.

➢ **Zucchini and Summer Squash:**

- Harvest when they are young and tender, usually around 6-8 inches long.

- Larger fruits can become tough and less flavorful.

➢ Beans (Green Beans, Snap Beans):

- Harvest when the beans are young and tender, before the seeds inside have fully developed.

- Pick beans regularly to encourage continuous production.

➤ **Peas:**

- Harvest peas when the pods are well-filled but still tender.

- Test by gently pressing your thumbnail into the pod. If it leaves an indentation, the peas are ready.

- **Onions and Garlic:**

- Harvest onions when the tops have fallen over and started to dry.

- Garlic is ready when the tops start to turn brown and fall over. Dig up a bulb to check for mature cloves.

2. Harvesting techniques to avoid damage to plants

➤ **Use sharp tools:** Sharp tools like pruners or scissors can make clean cuts, reducing damage to plants. Dull tools can crush stems and leave jagged edges, which may lead to disease or slower healing.

➢ **Harvest at the right time:** Harvest crops when they are at their peak ripeness. For fruits and vegetables, this often means harvesting when they are fully mature but before they become overripe. This ensures the best flavor and texture while minimizing stress on the plant.

➢ **Handle with care:** When harvesting fruits or vegetables, handle them gently to avoid bruising or damaging the plant. Support the stem or branch when picking to prevent tearing or breaking.

➢ **Avoid excessive trampling:** Try to minimize walking on the soil within the raised bed, as this can compact the soil and disrupt root systems. Designate pathways or install stepping stones to access different areas of the bed without compacting the soil around the plants.

➤ **Rotate harvesting:** Instead of harvesting all the produce from one area at once, rotate your harvesting locations. This allows the plants time to recover and continue producing throughout the growing season.

➤ **Prune strategically:** Regularly prune plants to remove dead or diseased foliage, as well as overcrowded or damaged stems. This not only improves air circulation and light penetration but also encourages new growth and healthier plants.

➤ **Mulch around plants:** Mulching the soil surface around plants helps to retain moisture, suppress weeds, and protect the soil from compaction and erosion. Use organic mulches like straw, shredded leaves, or compost to provide a protective layer around your plants.

> **Water properly:** Maintain consistent soil moisture levels by watering your raised bed garden appropriately. Avoid overwatering, as this can lead to root rot and other moisture-related issues, but also ensure that plants receive enough water, especially during hot or dry periods.

3. Post-harvest care and cleanup

> **Remove Plant Residues:** After harvesting, remove any leftover plant residues, including roots, stems, and leaves. These residues can harbor pests and diseases over the winter and may interfere with the soil's ability to breathe and absorb nutrients.

> **Compost or Dispose:** Compost healthy plant residues to create nutrient-rich compost for future use in your garden. Avoid composting diseased or

pest-infested plants, as this could spread the problem. Instead, dispose of them in the garbage or burn them if allowed in your area.

➢ **Mulch or Cover Crops:** Consider mulching your raised beds with organic materials like straw, leaves, or compost to protect the soil from erosion, conserve moisture, and add organic matter. Alternatively, plant cover crops such as clover, winter rye, or hairy vetch to enrich the soil, suppress weeds, and prevent nutrient leaching.

➢ **Soil Amendments:** Test your soil to determine its nutrient levels and pH balance. Based on the results, amend the soil with organic fertilizers, lime, or other soil conditioners to ensure optimal fertility and pH for the next growing season.

➢ **Till or Turn Soil:** If your raised beds allow for it, lightly till or turn the soil to incorporate organic matter, break up compacted soil, and expose pests to predators and weather extremes. Avoid excessive tilling, as it can disrupt soil structure and beneficial microbial communities.

➢ **Check for Pests and Diseases:** Inspect your raised beds for signs of pests or diseases. Remove any affected plants or debris and take preventive measures, such as rotating crops and practicing good sanitation, to minimize future problems.

➢ **Repair and Maintenance:** Take this opportunity to repair any damaged raised bed structures, such as loose boards or deteriorating supports. Clean and disinfect gardening tools to prevent the spread of diseases between plants.

➢ **Plan for the Next Season:** Use the downtime after harvest to plan your next crop rotation, order seeds or seedlings, and update your garden journal with notes on what worked well and areas for improvement.

4. Planning for future seasons and crop rotation

> **Plan Ahead:** Before each season, sketch out your raised beds and decide which crops to grow where. Consider factors like sunlight exposure, water needs, and the growth habits of different plants.

> **Crop Rotation:** Rotate crops annually to prevent nutrient depletion and minimize pest and disease pressure. Divide your crops into different plant families (e.g., nightshades, brassicas, legumes) and avoid planting the same family in the same bed two years in a row.

> **Soil Health:** Incorporate organic matter such as compost or aged manure into the soil between seasons to replenish nutrients and improve soil

structure. Cover crops like clover or vetch can also be planted during the off-season to prevent erosion and add nitrogen.

➤ **Succession Planting:** Plan for successive plantings to make the most of your space and extend your harvest throughout the season. After harvesting early crops like lettuce or radishes, follow up with heat-loving plants like tomatoes or peppers.

➤ **Companion Planting:** Take advantage of companion planting principles to enhance plant growth and deter pests. For example, planting basil near tomatoes can improve tomato flavor and repel pests like aphids.

➤ **Vertical Gardening:** Incorporate vertical gardening techniques to make efficient use of space. Trellises or stakes can support climbing plants like peas, beans, or cucumbers, freeing up ground space for other crops.

➢ **Record Keeping:** Keep a gardening journal to track which crops were planted where, their performance, and any issues encountered. This information will inform future planning and improve your gardening skills over time.